A NEW VIEW OF THE WORLD

A Handbook to

The World Map : Peters Projection

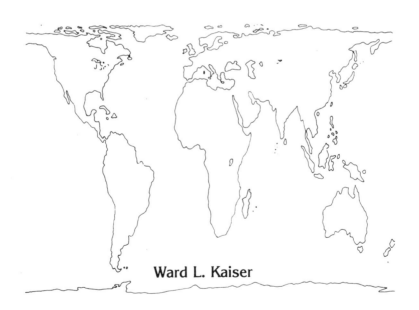

Ward L. Kaiser

Friendship Press • New York

Acknowledgments

Many people have helped shape the thinking and experience that this handbook represents. Particular thanks are expressed to Arno and Birgit Peters for the intellectual stimulus and personal warmth of their friendship; to Arthur Bauer for seeing the need for this handbook; to Marian Ziebell, whose comments have improved the text; to Professors Leonard Turner Guelke and Ralph R. Krueger of the Department of Geography, University of Waterloo, Ontario, for critical support in reading the manuscript; to the International Students and Scholars Program at New York University and to Bergen Community College, Paramus, New Jersey, for opportunities to lecture and teach in related fields. No person or organization listed here should, however, be held responsible for any of the views expressed.

Acknowledgment is also made to the following publishers for permission to use copyrighted material:

Buckminster Fuller Institute, Los Angeles, California: Dymaxion Map. For information about the Dymaxion Map contact The Buckminster Fuller Institute, 1743 S. La Cienega Blvd., Los Angeles, CA 90035.

Earth Observation Satellite Company, Lanham, Maryland: Landsat view of Los Angeles.

Library of Congress Cataloging-in-Publication Data

Kaiser, Ward L.,
    A new view of the world.

    Bibliography
    1. Peters projection (Cartography)   I. Title.
GA115.K35 1987   526.8'5   87-17607
ISBN 0-377-00175-9

Editorial Office: 475 Riverside Drive (772)
              New York, NY 10115

Distribution Office: PO Box 37844
              Cincinnati, OH 45222-0844

Printed in the United States of America

CORRECTION

The statement attributed to the American Congress on Surveying and Mapping on page 10, also referred to on page 31, was originally made in a bulletin of the Press and Information Office of the Government of the Federal Republic of Germany. The ACSM reprinted the material in its Bulletin No. 50 in November 1977 but has not made an official statement of its own on the Peters Projection.

CONTENTS

Fig. 1.
A view of the Los Angeles, California, area as seen by the Landsat automatic remote sensor.

Introduction: The Rediscovery of Planet Earth

This is an exciting time to be alive. We, the people of the world, are rediscovering Planet Earth. We're seeking new knowledge, just as Vasco da Gama and Henry Hudson and Ferdinand Magellan pursued fuller understandings of their world. We're also becoming aware that the *way* we look at the world is important.

In our ongoing search for knowledge we face possibilities never before available to human beings. Modern science and technology have opened new vistas. How many of us over 30, for example, would have believed we would see the time when the circumference of the entire earth could be digitally photographed and mapped in just 18 days? Yet that is precisely what Landsat, an automatic remote sensor operating from space, is capable of doing. (See Figure 1.) Our era can provide us with more accurate, more reliable data in greater quantity than previous generations dreamed of.

We are, at the same time, testing new ways of looking at the world, new perceptions of what the mathematical and scientific realities mean. We have come to recognize that people first shape their maps, then their maps shape them. We owe it to ourselves to be aware of the messages maps send, whether subtle or blatant, and to judge those messages from a consciously held system of values.

This book is, as the subtitle says, "A Handbook to the World Map: Peters Projection." It will refer to other maps, but it will focus clearly on this one map of the world, sometimes described as "the map for our time." It will point to the scientific and mathematical basis for map making, and deal with questions of meaning and perception.

We live in the Age of Science. We have learned to test

products and evaluate claims on the basis of objective standards. We are coming to realize that our images of our planet often lack objectivity. They are not as accurate as we have asumed. In our day, however, a new possibility has come into the picture: a world map that achieves a level of accuracy never before thought possible.

At the same time, questions of perception may loom even larger than questions of scientific accuracy. From the time of its introduction in Europe in 1974, the Peters map has elicited both support and criticism. Dr. Peters comments:

> . . .public discussion was such as had not been known in the history of cartography. I attribute this to the fact that the debate over my map was in reality not a struggle about a projection as such but over a world picture. Clearly, ideology had entered the struggle.[1]

This book is written, then, to share with you the exciting story of the Peters map, setting forth the related scientific questions and the attitudinal adjustments we face. You will, of course, make up your own mind about the map and the questions it raises. And, for those who teach or lead groups, we provide suggestions for using the map in class.

Welcome to the human adventure of rediscovering Planet Earth!

1.

Maps and What They Represent

Suppose you want to order a world map. You may wish to show where your company has business contacts; you may want it for classroom teaching; it may be for purposeful decoration on a wall of your home, or as a reference to consult when you follow world news. Since there are many maps available, how do you go about choosing the one that is right for your purposes?

Leaving aside for the moment such variables as size and price, ease of ordering and whether the colors of the map blend in with the chromatic scheme of your room, let's talk about the map itself.

The first thing to know is that there is no absolutely accurate map of the world. There never has been. There *cannot* be. To see why this is true, think of an orange or a tennis ball. Visualize scoring the skin and peeling it off. One thing is sure: what you get won't lie flat. Even if you did manage somehow to flatten it, shrinking it here, stretching it there, the final result would be so deformed that you'd have a hard time ever turning it into a tennis ball or an orange skin again.

That suggests the first problem cartographers face. How can the surface of a sphere—our planet[2]—be transferred to a plane—a flat surface? Transferring the surface of one sphere to another sphere, while requiring considerable expertise, presents no insoluble problem. A small earth-globe can provide an accurate representation of the planet. Only the scale is reduced. But translating from a three-dimensional world to a two-dimensional map, on the other hand, means accepting some compromises. The question becomes: Which ones?

DESIRABLE CHARACTERISTICS

To be more specific, let's ask: What would an ideal map look like? We can identify certain desirable characterisics or properties. These would certainly include:

Equality of Area

On a map with this quality any area—a country, for example—could be compared with any other area, with advance assurance of accuracy. There would be no distortion of size, either through enlargement or shrinking of any area, land or water.

Fidelity of Axis

The common convention that North-South (axis) lines are shown as verticals is very useful. For one thing, this corresponds with our normal sense of direction, in which North feels "up" and is shown at the top of the map. Historically, North has not always been given "top billing," and there is no scientific reason why it should be. Nevertheless, until a better system is devised, both makers and users of maps find it most appropriate to place North at the top. When any map takes the further step of making all North-South lines run parallel to each other, we say that map has fidelity of axis. It may also be described as "user-friendly," in that it provides strong support to those who want to "get their bearings" or orient themselves.

Fidelity of Position

Remember how a compass is set up? The East-West line invariably intersects the North-South axis at right angles. On an ideal map, then, this right-angled pattern would be maintained. The result would be a rectangular grid. On such a grid, when all East-West lines are parallel, users can see at a glance distance north or south of the equator. Position, or the point where lines of latitude and longitude intersect, is clear. Users have certain basic information about climate. On any given day, the sun will strike all places on any East-West line at exactly the same angle. Note also that in the real world all East-West lines run parallel, with the same distance between them through their entire length, hence the term *parallels* of latitude.

Equality of Linear Scale

Accustomed as we are to reading road maps, with their reliable measurements of distance in miles or kilometers, we tend to expect the same on a world map. On an ideal world map, we might say, we could confidently measure distance from any Point A to any Point B.

Accuracy of Shape

Land masses and ocean areas have contours that give them characteristic shape; an ideal map would represent these with recognizable accuracy.

Fidelity of Angle

Perfection in a world map would surely mean also that the user could set a compass bearing from any point A to any point B and know that cartographic reality matched geographic reality: That the map and the world gave the same readings. During the Age of Discovery, when European sailors explored the expanding world, this was of great value. Today its practical use is sharply diminished, though the quality itself rightly belongs on a list of theoretical properties that a perfect world map would possess.

You may think of other desirable characteristics. Would you include having the equator at midpoint? How important is it to you that a map offer a carefully thought-through color scheme? What land mass—or ocean—would you like to see at the center? Clearly, the list could be extended. Nevertheless, the six cartographic and mathematical qualities we have listed are of prime concern, and we so treat them here.

Column A	Column B
Area	Linear Scale
Axis	Shape
Position	Angle

Has any map achieved all six of these qualities? Alas, no. The mapmaker's predicament may be seen as a variant of that faced by a group of people in a Chinese restaurant: They are not at liberty to order on the basis of preference alone. Their options are structured, or limited, by the offerings in each column of the menu. In this instance, those map qualities listed in Column A have been generally considered to be incompatible with those in B. To be more specific, the mapmaker has had to choose between fidelity of area (Column A) and fidelity of angle (Column B). These two properties, according to the common wisdom among cartographers, cannot be combined in one map.

In our day, however, a new and different approach has become possible. Dr. Arno Peters claims to have broken out of these very limitations. Whether one sets out six cartographic qualities as we have done here, or ten or twelve as may be done, the Peters Projection principle now makes it possible to achieve all of them to an extent never before accomplished.[3] This is, in fact, the accomplishment that ushers in "the new cartography," sharply contrasted with the old.

Across the centuries different mapmakers of the "old cartography" have chosen different priorities, as you might expect. No one has achieved them all; the perfect map has not existed. It is useful to evaluate any map, therefore, from the standpoint of these qualities. Consider the following examples:

The Mercator

—First developed in 1569 for European navigators.
—Named for its creator, cartographic scholar Gerhard Kremer,

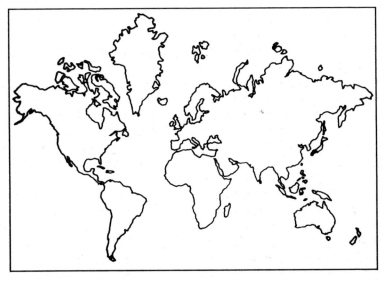

Fig. 2.
The familiar Mercator map, a scientific breakthrough in 1569 but outdated today.

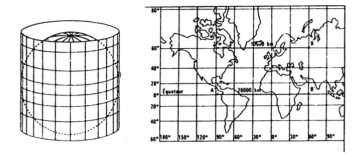

Fig. 3.
To get a rough picture of one of the projection methods used, imagine a light bulb inside a transparent earth, projecting the earth's features into the inside of cylinder wrapped around it. When the cylinder is unrolled the sphere's features are displayed on a flat surface.

whose surname (meaning "merchant" in English) becomes Mercator in Latin.

—As the first precisely calculated world map, represented a major accomplishment in cartography.

—Based on the principle of cylindrical projection. See Figure 3.

—Fails the equal area test, since it seriously distorts size. See Figures 4 and 5 as examples.

—Has fidelity of axis and of position. On these two important properties it achieved a high degree of usefulness in the age of European expansion, as adventurers set out to explore and exploit the world.

—Often interpreted as having fidelity of angle. Dr. Peters has, however, demonstrated that it fails this test. No two-dimensional map, he points out, can claim this quality. Therefore fidelity of angle is not a meaningful category.[4]

—Progressively deforms shape, moving from zero deformation (i.e., complete accuracy) at the equator to absolute distortion at the poles. Mercator himself recognized this problem as inherent in his projection system, and wrote, "The map cannot be extended to the Poles themselves as the degrees of latitude finally increase to infinity."[5]

—Places the equator below midpoint—about two-thirds of the way down.

—Has been reworked many times in the four centuries since Mercator's day, as new areas have been discovered (Australia and Antarctica for example) and as cartographers have come to grips with its limitations.

—In spite of such limitations, still dominates the market.

—Remains highly influential in "shaping" people's views of the world. Most of us, if asked to draw a map of the world from memory, would roughly reproduce the Mercator projection.

The Van der Grinten

—Developed in 1904.

—Is one of several map projections offered by the United States Geological Survey and the National Geographic Society. The *National Geographic* Magazine inserted this map into its December 1981 issue, for example.

—The National Geographic version provides a scale of distance, which certainly makes it rather special among world maps. (Nevertheless, this feature is of questionable value. For example, if you want to know the distance between Chicago and Paris you would, typically, come up with a figure around 5900 miles, based on this map's scale. Airlines calculated the distance as 4133 miles[6] — that is, the map shows a discrepancy of 42.77 percent! Actually, anyone with, say, a college major in geography knows that airlines fly the great circle[7] route, which is shorter than a flat-map straight line. Yet even if we take highway distances rather than airline routes as the standard, the scale provided on this map proves misleading. Measuring the distance from Dallas to Chicago yields a value of about 1070 miles; road atlases generally give this at about 790 miles.[8]

—Does not have equality of area.

—Lacks fidelity of axis.

—Lacks fidelity of position.

—Places equator below the middle of the map. Thus, like Mercator, it gives unequal prominence to the Northern Hemisphere.

—Viewed by some as inferior to the Mercator, in light of the characteristics just outlined.

"Spaceship Earth" (Dymaxion World)

—Invented by the late Buckminster Fuller, who also created the geodesic dome.

—More formally known as Buckminster Fuller's Dymaxion Sky-Ocean World.
—Published March 1, 1943 in *Life,* though earlier versions date to 1927, when Fuller privately published his book *4-D.*
—Enjoys equality of area or at least a close approximation.
—Lacks vertical axis.
—Lacks fidelity of position.
—Like most world maps, does not provide a scale. Nevertheless, for the sophisticated user, distance along great circle arc segments[9] can be accurately determined.
—Faithfully represents land mass contours.
—In part because it "seems strange" to the eye and the brain, in part because of its broken surface and sharp angles, it has never achieved a high level of public acceptance.

The Peters Projection

—First published in 1974 in Germany; first English-language version, 1983.
—Created by Dr. Arno Peters, the German historian whose life work emphasizes the equal status of all peoples.
—Drawn not for navigators of wooden sailing ships nor for modern military strategists but for the people of the world.
—Accurately represents all areas according to relative size. One square inch anywhere on the map represents a constant number of square miles[10]
—Has fidelity of axis.
—Has fidelity of position.
—Includes some shape distortion. To many (including this writer) this can lead to a negative first impression. In point of fact, however, this deformation is objectively less than is found on traditional maps. It is just that we have grown so accustomed to *those* relational distortions that we actually suppose they represent reality!
—Has called forth some criticism. Sometimes this centers on the question of shape. Perhaps more often, negative feelings arise out of subtle sense of being threatened, which we will discuss more fully in the next chapter.
—Enjoys remarkable acceptance to date. Volume users include agencies of the United Nations, schools and colleges, churches, Third World action groups and Peace Corps Volunteers.
Support for Professor Peters' map has been forthcoming

from a number of professional communities. Geographers and cartographers are among these. Thus the American Congress on Surveying and Mapping could say:

[Dr. Peters' map] shows all countries and continents in their correct relative proportions. This absolutely area-factual Peters Projection furthermore keeps the unavoidable distortions of forms, distances, and angles so minimal that a world picture of great faithfulness to reality came into being.[11]

The late Professor Carl Troll, former president of the International Geographers Union, called the Peters map "the best I know."[12] Never before, he pointed out, had so many desirable qualities been united in one map.

From a Third World vantage point, the University of the West Indies, comes another strong statement of support. Says Dr. Vernon Mulcansingh, Chair of the Department of Geography:

[Dr. Peters'] crowning achievement. . . represents a burst of brilliance that can be compared with any major breakthrough in any field of science... For the first time in history, almost, we are seeing on paper what our world really looks like.[13]

Still, it is one thing for the community of scholars to recognize high attainment within their own field of expertise. The rest of the world typically goes on, unmindful and unaffected. In this instance, however, the world is paying considerable attention. How do we explain that? What lies behind the fact that many who have never shown much interest in geography. . . or mathematics. . . or even in map reading if they were ever in Scouting. . . now confess to a great excitement about this particular map? Additional help in answering that question will be provided in the chapters that follow.

2.

The Messages Maps Send

Maps send powerful messages. If a picture is worth a thousand words, a map of the world is worth a thousand pictures as we try to understand our place in that world.

The Mercator map is a prime example. Let's examine some of the messages it sends. Draw diagonal lines from corner to corner across its face and what do you find at the center of the world? Logically, the center could be anywhere on the equator—Ecuador, Kenya, Indonesia, for example—since the equator is, by definition, the imaginary line that separates the world into equal halves: Northern and Southern hemispheres. In this instance, however, you find the center to be western Europe. Why? Why should Germany, Gerhard Kremer's adopting land,[14] lie close to the visual center — as if it were situated next to the earth's belt when in fact it belongs in the top one-quarter of the earth's surface?

The point is not that Mercator deliberately falsified the picture. He was following almost universal precedent in setting his own land at the center of everything. This had, in fact, a positive effect in this sense: It facilitated the use of his map by European navigators in the age of exploration.

We, however, are not sixteenth-century European sailors. For us the continued widespread use of the Mercator as a general-purpose map presents serious problems. We are neither fair to Kremer nor acting in our own best interests when we force his map to function beyond its originally intended use or its capabilities. Specifically, we have tried to turn a navigator's tool into a teaching tool—a most unhelpful decision, for which Mercator is certainly not responsible. The Mercator map, as it is most often used today, lends support

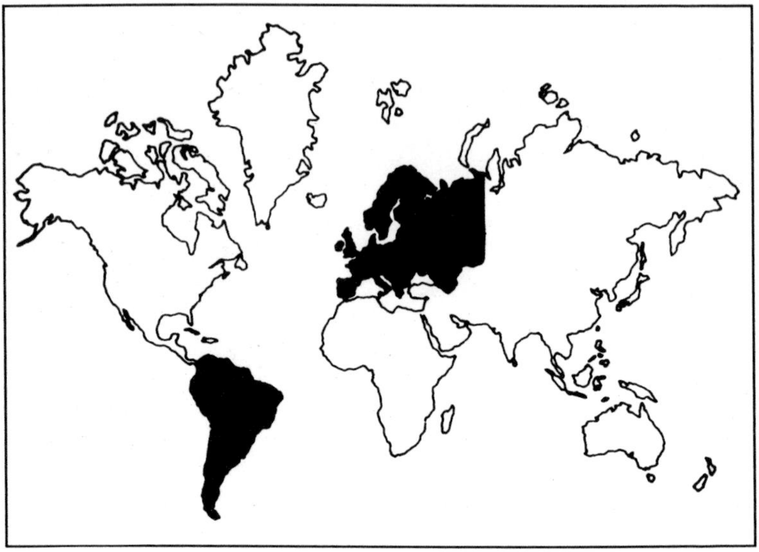

Fig. 4.
*The Mercator map makes Europe look larger than South America;
however, it has only 3.8 million square miles compared to South
America's 6.9 million square miles.*

to the assumption—unfortunately all too common even now
— that nationalism or ethnocentrism or even racism is all right:
that it is grounded in geographical realities. Please read these
words carefully: I do not assert that this map is the source
of the problem, but that it lends support to the problem. It
does so, not out of any original evil intention, but precisely
because many people accept it uncritically. To the extent that
people believe it to represent "truth" or to be "accurate,"
to that extent it is inextricably linked to a major, worldwide
problem of our time in history.

Ethnocentrism or racism is the belief that one's own people or race are superior to all others. To see it in its extreme
form, look at apartheid in South Africa. Less obvious but still
significant forms of the problem exist all around us and within
us. The sources of this dangerous disease are many, but surely
one of its subtle supports is the pervasive use of a map that,
in spite of all known facts:
—enlarges those areas of the world historically inhabited by
whites,
—shifts those same areas to the heart and center of the world's
stage, where they do not belong, and

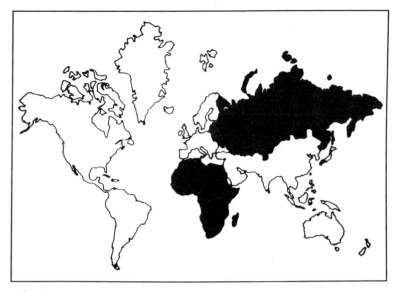

Fig. 5.
Africa, with its 11.6 million square miles, is much larger than the Soviet Union at 8.7 million. But who would know that from this map?

—minimizes the importance of what we think of as "the South," including most of what we designate the Third World.

Those who use the Mercator for purposes other than steering a sailboat across the sea may be innocent victims of ethnocentric or racial bias. They may be — again, quite unconsciously — among the perpetrators of such bias. At very least they lend unwitting support to a view of the world that neither modern science nor mathematics nor the world's great religions nor common sense can justify.

How then does it happen that many still use that four-centuries-old map? After all, we constantly confront it in classrooms. And in textbooks. And as background for world news presentations on television. We even see it displayed behind the United States Secretary of State when he interprets foreign policy.

Why is Mercator so widely used? Since no one can provide a fully satisfying, verifiable answer, anyone is entitled to an educated guess. The possibilities are mind-boggling. Consider educational settings first. Is it possible that textbook authors and classroom instructors all grew up on the Mercator, so that they instinctively turn to it when they need a map today? Does the problem lie with the publishing community not having

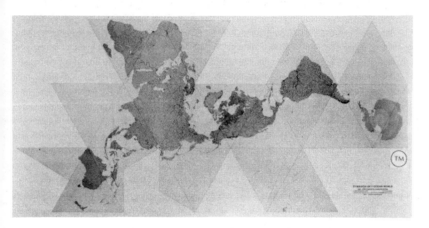

Fig. 6.
Buckminster Fuller's Dymaxion World Map projection.

made good alternatives available in classroom size? As for TV, maybe the poverty-level budgets on which networks operate cannot handle the cost of replacing existing stage sets.

In any case, the map we use in all these various settings serves both as a mirror of our mindset and as an educational device that helps to shape the future.

Is it too much to hope that when we look in that mirror we may not be embarrrassed to find what is in our minds, and that when we evaluate the educational devices we use we may find them moving us toward a positive world future instead of a colonial past?

While people must dig out the messages Mercator sends, the message of certain other maps is more accessible. An example is provided in Figure 8. This cartogram or map carries a highly specalized purpose: to show how much the major nations of the world spend on armaments. For such an objective, physical contours are irrelevant. It is inconceivable that anyone would use this map to sail the seas or compare India and China for phyical size or to find out which is closer to the North Pole, Ottawa or Moscow. (Yet we often press maps to serve purposes beyond their competence!) If we will honor the single purpose and clear message of this cartogram, it can be very useful.

Buckminster Fuller's Dymaxion World Map sends its

A New View of the World

Handbook to the World Map: Peters Projection

messages. Since there are several ways to reassemble this flat map into a multisided "globe," it is not surprising that each approach carries its own message. Let Fuller himself interpret two of these:

> ...one of these pictures is...the One Ocean World, fringed by the shoreline fragments.... It discloses the relative vastness of the Pacific and emphasizes that ocean's longest axis, from Cape Horn to the Aleutians. Oriented about the Antarctic, the waters of the Indian and Atlantic Oceans open out directly from the Pacific as lesser gulfs of the one ocean....
>
> ...compare the impressions derived from looking first at the one-continent arrangement and then at the one-ocean assembly. Turning away and reporting his [sic] impressions...[the reader] would be inclined to testify that these maps were composed of different components; that the one-continent map was composed of seventy-five per cent dry land area, that the one-ocean map was comprised of ninety per cent water area.[15]

Either way the messsage is striking: The world is one vast ocean punctuated by chunks of land, outcroppings of the world's

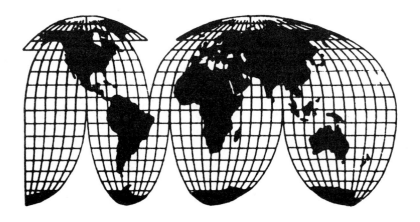

Fig. 7.
A Homolosine World Map Projection.

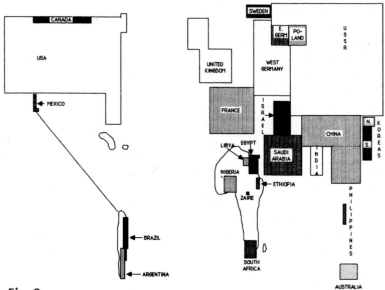

Fig. 8.
A cartogram or quantified-information map. This one, created on a computer, sets forth size not by land area but by military expenditures.

one continent, or one vast continent set in a global sea.

Fuller also saw his invention as a long-needed corrective to an outdated perspective on the world. He spoke of

> . . . our prevailing public ignorance of dynamic Air-ocean geography, we being as yet blinded by old-plate-stocked Mercator projection vendors, and by the historical east-west orientation inertia. It is hoped that a new north-south dynamic world orientation will be aided by the *Dymaxion Airocean World.*[16]

In this respect it is significant to note that both Fuller and Peters speak of the North-South reality. They see it as a new and important factor in world history, which the old maps do not sufficiently recognize. Their purposes are divergent, however; Fuller was largely concerned with helping the United States achieve its potential, through the use of creative imagination and forward-looking technology; Peters is more clearly focused on justice for all people, recognizing the values and contributions that all nations and all cultures can bring to the emerging world civilization.

Even the colors used on maps convey important information and send messages. At one time it could be claimed that "the sun never sets on the British Empire." Accordingly it was appropriate and helpful to depict that empire in a single color—typically red. Now that the empire has passed into history, the tradition—surprisingly — lives on in some color schemes. India, Canada and Nigeria among others are thus tied to their history rather than to present reality.

Most maps, however, have moved beyond that problem in the use of color. The National Geographic's version of Van der Grinten, for example, shows all land masses in white, with only national borders getting bands of color. Color gradations are reserved for the oceans: the deeper the blue, the greater the depth. The continent of Antarctica is an exception: it gets no colored border, as is perhaps fitting for so icy an area.

Another map with a clear message comes from Australia, as seen in Figure 9. Call it upside down if you will, but first consider the official preface to the map, which calls it

...the first step in the long overdue crusade to elevate our glorious but neglected nation from the gloomy depths of anonymity in the world power

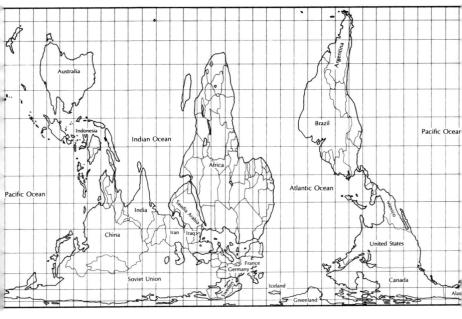

Fig. 9.
What's "up" in the world? Has this map been printed upside down? Is Australia really "down under"—or have its people offered a corrective to our Northern bias?

struggle to its rightful position — towering over its Northern neighbors, reigning splendidly at the helm of the universe. Never again to suffer the perpetual onslaught of Down Under jokes—implications from Northern nations that the height of a country's prestige is determined by its equivalent spatial location on a conventional map of the world. . . . Finally, South emerges on top.[17]

Viewed another way, this map makes it very clear that any map provides more than geographical information. Beyond the obvious there is the subtle; beyond the facts there lies the whole realm of meaning; in addition to pure science there exists the fruitful world of what we call messages. Values and images—images of one's own people and of the other peoples of the world—are born, live and die here. Long ignored by professional cartographers as a subject outside their jurisdiction, this is a concern that merits attention by all the people. It is too important to be left to the professionals.

The Peters map also communicates its special message. Elements of that message may be identified:

—Peters takes the world seriously. This map says, in effect, that the world is so important a place that we ought not be content with outmoded, inferior and distorting ways of depicting it.

—Peters takes modern science seriously. It incorporates advances in human knowledge of the earth and in mathematics, enhancing the reliability of its world picture.

—Peters takes the world's people seriously. Far from exalting a (white) minority while relegating other peoples to relative obscurity, it provides every nation its rightful place.

—Peters takes the user seriously by giving the essential data, comparisons and interpretation on which to base an independent judgment. It holds out the promise that with the aid of this learning tool all of us who share this planet may yet find liberation from the domination of a distorted and distorting, out-of-date, colonialist view of the world.

3.

Questions People Ask

Thus far we have provided an overview of the objective or scientific basis of cartography and have opened up the question of the messages maps convey. We have looked at a variety of available maps as background for appreciating the distinctive characteristics of the Peters Projection.

It may well be, however, that you have questions in your mind about the Peters world map that have not been touched upon. In this chapter I'd like to try to respond to those. Since a book is not an interactive medium, I cannot receive individual questions from you, the reader, and respond; what I can do is identify those cogent questions that have been asked since the introduction of this map to the English-speaking world. I take these, largely verbatim, from radio and television interviews, lectures, workshops, customer queries and personal conversations.

In this chapter, then, you may find questions you would like to ask, along with responses that we trust will shed some light. Or, if you are called on at some time to interpret the Peters map, these questions may help prepare you for that opportunity.

What is the thinking behind the choice of colors on the Peters map?

Colors are an important part of the language of maps. We did indicate that sometimes color is used to provide political information, as when India was printed in red to show it belonged to Great Britain. Sometimes color can be used to make a political statement. Depending on the color chosen for the Malvinas/Falkland Islands or for certain disputed areas

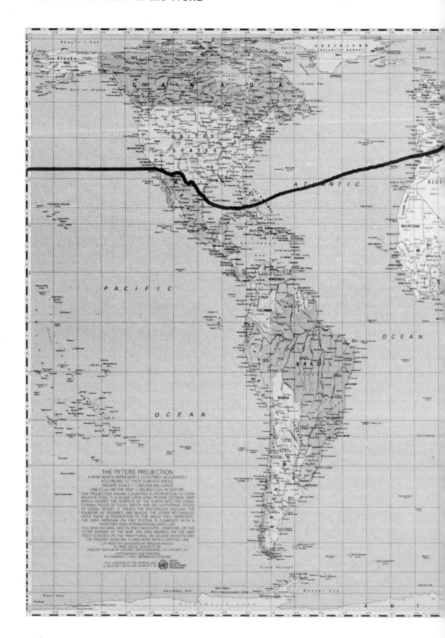

Fig. 10
The Peters Projection of the World Map, because it is area-accurate,
can be used to show modern economic and geopolitical facts in a
realistic way. On this map "North" (above the line) represents 31

percent of the world's land occupied by 25 percent of its people.
"South" (below the line) represents 69 percent of the land mass and
75 percent of its people.

in the Middle East, the cartographer or publisher might well be making a statement of political preference.

The Peters world map sets aside the colonial connections of the past in favor of the realities of the present. One of those realities is the heightened sense of identity among the peoples of the world, particularly in Asia, Africa and Latin America. Regional and national identities more and more take precedence over a relationship that owes its origin and its continuation to forcible conquest and foreign domination. Therefore Peters conceived the idea of showing a whole region in one dominant color-family, with each nation having its own variant. Thus the "family connections" as well as the separateness of each country can be shown.

To my knowledge there is no other world map that uses color in precisely this way, or that takes regional awareness so seriously.

This map would be more fully accepted if it didn't look so strange. Some even say it reminds them of Salvador Dali's paintings: you know, watches melting over the side of a table and all that. Couldn't it have been made less different?

Can you help me know—perhaps with some examples — what you mean by the term "less different"?

Take Africa as an example. Couldn't it have been depicted more normal—not so long and narrow?

Any mapmaker lives within certain constraints. First of all, there are the requirements imposed by geography itself. Any land mass must be shown accurately in terms of its coordinates, that is, those points of intersection of latitude and longitude.

Beyond that, Peters chose to incorporate into his projection system the cartographic properties of fidelity of axis and fidelity of position that we spoke about in Chapter 1. This meant that he was working with a rectangular grid. He also set out to achieve fidelity of area. Taken together, these goals impose extremely strict requirements on the mapmaker. He or she cannot "fatten" Africa's midsection at will, nor reduce its long stretch top to bottom.

There is, then, no way to play with the picture to make it "less different" without sacrificing at least one of the three fidelities: area, axis and position. Yet it is precisely these properties that make this a very good map.

What are some ways people have used this map?

There are some fascinating stories here. The General Board of Global Ministries of The United Methodist Church has a Peters world map etched in glass at the entrance to its New York office. The result stands six feet high; visitors cannot miss it as they come in. This serves as a fitting reminder of that agency's concern for the whole world, a world in which justice and equality will be fully known by all the people of the world.

The president of a major multinational corporation sent his chauffeur to the publisher's office after normal closing time to get a copy so he would have it ready for display at his next board meeting.

An interior decorator learned of the map through a design magazine; liking it, she recommended that it be hand-painted on the wall of a forward-looking client's family room.

A high school teacher is using the Peters map to help his students gain a more accurate view of the world they live in, and at the same time to develop a positive, questioning attitude toward life. "These days," he says, "young people have so much thrown at them from all sides, that unless they learn to sort it out and do some critical thinking for themselves, they're lost. Just because they've always seen a Mercator on the classroom wall doesn't mean that's the real world. I find we can take the discussion from there to other questions where they need to exercise some independent judgment. Just because some media idol says it's OK to do drugs doesn't make it harmless; just because the majority of their friends do something, doesn't make it right. This map can lead to some amazing discussions!"

Most people have less dramatic stories. They would tell of personal conversations about the map and the world it depicts, putting it on the wall of their home, of using it in college courses in geography or in informal educational settings such as world mission studies in churches. And of course Third World and development action groups are major users.

This map is in use in the Vatican and the offices of the World Council of Churches. NATO forces have used it. United Nations agencies are among the strongest supporters of the Peters projection.

Is this map available in other languages?

It is now available in six languages. In addition to German and English, it can be had in French, Dutch, Spanish and Italian.

Some people say this map is communist-inspired. Is it?

If it were, wouldn't you expect that it would by now have become available in the Soviet Union and China? It has not been published there, however. Nor in East Germany nor any other country in the Soviet bloc. Not yet, anyway.

Also, does it not seem strange that a "communist" map actually cuts the USSR down to its proper size? It is no longer the gigantic presence that it appears on traditional maps.

Finally, the objective basis of this map is clear and open for all to see. It is founded on universal principles of mathematics. Clearly, those principles favor neither communists nor capitalists; they simply reflect the truth about our world.

You mentioned controversy. What are some of the criticisms that have been voiced concerning this map?

A few people have objected to using the Mercator map as a basis of comparison, as is done through the inset maps on the face of the popular version of the Peters map. In essence they are saying that Mercator was never intended to provide a definitive picture of the world or to teach geography. It was to be used *only* for navigation.

These people tend to be professional geographers or cartographers. They form the elite of the profession. They enjoy a level of sophistication that few of us will ever achieve in a lifetime of study and work. But one wonders if they have looked lately at how often the maps we common people turn to are Mercators.

A second criticism comes from an opposite point of view. It says that we should not be so critical of the Mercator. We *better* take it seriously. It has stood the test of time. Since it is used more widely than any other map—in classrooms, for wall maps, on TV screens and so on—it has got to be good. And since it is good, why do we need to replace it?

Obviously these two groups cannot both be right. Their criticisms come close to cancelling each other out. Stated another way: I agree that Mercator was never intended as a teaching tool. I assert that it should not be used that way. Truth is, however, it *is* used that way—and very extensively. When I hear that we have taken Kremer's original purpose seriously, and stopped forcing his creation to do what it cannot do and should not do. . .when I hear that all Mercator maps have been removed from classrooms, from TV shows and from federal government offices, I'll know that we have taken an important step forward.

As for the statement that Mercator is the most popular projection and must therefore be best, I doubt that value in this instance can be determined by votes. No matter how many people suppose the Soviet Union is larger than Africa, the facts do not support thém. No matter how many people have lived with Mercator maps all their lives, it is still not a good map for most purposes.

A third criticism says that the Peters projection is to be rejected because it has shape distortion. When that comment comes from a lay person, I try to respond as I did in Chapter 1, pointing out that mapmakers face choices. They cannot attain all properties. Dr. Peters has chosen what we called "Column A" goals—fidelity of area, fidelity of axis and fidelity of position—rather than absolute fidelity of shape. I firmly believe that is a wise choice, and his map is eminently useful because of that.

A word of caution here, however. When we talk about distortion of shape, distorted compared with what? If—as often is true—we mean as compared with the map in our minds, typically Mercator, forget it. That map is fatally flawed, and gives us no standard for judging shape and size. It was never intended to do that.

Moreover, many maps in common use deform shape. Show me a map that has fidelity of shape and I'll show you a map that introduces other, more serious distortions.

Let us also note that in the case of the Peters projection this deformation is spread evenly over the face of the map, and is kept to a minimum. Whereas Mercator is accurate only on the equator and begins immediately to distort both north and south of that line, approaching total distortion at the poles, Peters achieves accuracy at 45 degrees latitude north and south. Maximum distortion is therefore, at one stroke, reduced by half. At the same time, Peters achieves proportionality, by which we mean spreading the necessary error out evenly across the whole map.

Occasionally even a professional geographer will point out that the Peters Projection has shape distortion. But surely such distortion should not surprise us! We know that some distortion of shape is inevitable if three fundamental properties are to be attained: fidelity of area, fidelity of axis and fidelity of position. More realistic is the position taken by the authors of a standard geography text, who say, "... although shapes of small areas will be good on conformal projections, the shapes of large areas will be deformed, *as they are on all projections.*"[18] These same authors also point out that "relative area is of fundamental importance."[19]

What do people who have used this map think about it? What have they said?

Judging by the comments I have heard, people are having a very good experience overall.

First, the North American source for quantity orders, Friendship Press, is enjoying many repeat orders. It's hard to quarrel with that. To date the map has sold more than 16 million copies worldwide.

Often customers write unsolicited letters of commendation for making this tool for world understanding available. Similarly, following a workshop on using the Peters world map in global education, the following were among the evaluations:

> This workshop helped me see how my understanding of the world has been distorted and inadequate. I'm glad to have it replaced by a more correct one, especially to know that this one supports fairness to all peoples.

> As you know, I came to this class with strong negative feeling about the Peters projection. I once studied geography, and took a course in cartography. I frankly didn't like what I saw and heard about this new map. I have now changed my mind. From now on this map only will be on my wall.

Organizational users likewise are enthusiastic. Arthur O.F. Bauer, a staff member of the Lutheran Church in America, has distributed many thousands of copies among his constituency. He says:

> The Christian message includes an emphasis on justice for all, based on the love of God which is extended to all. That message cannot utilize a map that sets forth an inaccurate, distorted world view. The Peters map appears to be the best education tool for showing us our place on earth. The values and purposes of this map coincide well with the teachings of the Bible and the church.[20]

In a number of localities there are citizen and teacher groups seeking to have the Peters map introduced in their school systems. They have experienced what it can mean in developing realistic concepts of the world, and they covet that enlargement of vision for all, particularly for students. One such example is Texas, where Broader Perspectives, Inc., a nonprofit organization concerned with education, presented its testimony before the State Board of Education. In March 1987 they reported that the Board had accepted 90 percent of their recommendations. These included changes that will affect all geography texts to be used in Texas in 1988 and beyond. Elizabeth Judge, executive director of the organization, reported that

> For world geography, texts now must compare map projections for accuracy of area, distance, shape, and direction. At the hearing on content requirements.... [I] compared Mercator's projection to the new Peters projection. Mercator's projection, which is most commonly used, exaggerates the size of the nations in the northern hemisphere; it was designed in 1569 for sea navigation by the colonial powers. The Peters projection shows the true area relation between continents of the northern and southern hemispheres. Since society places much emphasis on size, the Peters projection demonstrates more accurate and objective perceptions of the significance of nations in both hemispheres. Africa and South America finally have accurate size representation.

The significance of the Texas decision is wide-ranging, since most textbook publishers will sell the same text nationwide.

Oxfam America, highly regarded world relief agency, offers the Peters map to its supporters as "the most accurate map of the earth's surface yet drawn." And Harper's magazine called it "the first honest map of the world."

It seems to me that if we want the most accurate portrayal of our world we have it in a globe. Why don't we just all use globes and forget maps altogether?

You are right in your statement that a globe, and only a globe, can be accurate. Both shape and size can be correctly reproduced on it, for example.

Nevertheless, a globe has severe built-in limitations. For one, how would a teacher use a globe in teaching a class? To enable all students to examine it at the same time one would need to build a room around the globe, since so large a visual aid would never fit through an ordinary doorway! Even then, students would see less than half the world at any one time. Its equal-area advantage would be lost if students could not see, at the same time and on an equal plane, the two or more areas to be compared. Globes have a portability problem as well. So the list of inherent deficiences continues. . . .

Flat maps are clearly needed. The Peters Projection meets our human hunger to understand our world with precision and clarity.

4.

Teaching a New World Vision

This handbook takes as its purpose providing essential information suggesting how to deal not only with the *what* of *maps* but the *how* of *groups*. In this chapter we turn especially to the how.

Situations and needs will vary. Not only will they vary from person to person; they also will change from time to time. You may be a teacher of geography or social studies in a position to use or recommend helps for the classroom. You may be a parent who, wanting to nurture your family within a realistic and appropriate world view, seeks out ways of enlarging vision and understanding. You may belong to a voluntary organization such as a women's group in a church or synagogue, alert to the need for stimulating material to develop world-mindedness. You may be part of a community group concerned with Central America or race relations or peace action. You may be an individual with enough intellectual integrity that you want to pursue this subject further for your own enlightenment. However you see yourself, there will be, we trust, some ideas here to help you. But because of the broad range of possibilities to be covered, this section cannot provide detailed instructions for all possible situations. You will need to tailor the ideas according to your need. Let your creative imagination come into play!

The process we set out here is presented in four "phases." In a classroom or informal educational setting you will find them readily fitting into four sessions. They are:

Phase One: The Science of Making and Reading Maps
Phase Two: Exploring Map Messages
Phase Three: Maps Through History
Phase Four: Toward a New View of the World.

PHASE ONE: THE SCIENCE OF MAKING
AND READING MAPS

If yours is an informal group or classroom setting, display a variety of maps. Sources may include members' homes, schools and colleges, public libraries, book and map stores, world development agencies and religious institutions. Do not overlook those in encyclopedias. At very least you should have the Mercator and Peters maps. You will do well to have Van der Grinten, Fuller's Dymaxion, a homolosine map (the kind that looks like the skin of a peeled orange) and possibly others also. Try to have a globe of the earth. Having a copy of *The New State of the World Atlas* (see Resources) would also expand the possibilities. Let participants compare and contrast these. With one person responsible for keeping notes, let the group develop lists as follows:

1. What are the distinguishing features of each map?
2. What similarities are there among them?
    A. Features common to all;
    B. Features common to two or more.
3. What are the significant differences?
4. What do you suppose was the mapmaker's purpose in preparing each map?

Next let the class or group develop a list of desirable characteristics of maps. What would they expect to find in an ideal map? When the group has completed such a list—with or without achieving consensus—refer to the material in Chapter 1 on map properties. Explain the choices between Column A and Column B. Which map qualities can class members discover in the maps before them?

To check their own answers, let students turn to the section of Chapter 1 which analyzes particular maps according to their most significant characteristics. (If your group is ready to go beyond this level of analysis, you will find a more extended treatment in *The New Cartography*, pp. 105 – 118. (See Resource List.)

Stress the importance of the scientific advance that the Peters map represents. It was the first to achieve to such a remarkable degree the three properties of Equal Area, Fidelity of Axis and Fidelity of Position. Others had approximated these, or had attained one or two. Thus the mathematical or scientific superiority of this projection becomes apparent. You may wish to quote statements that recognize this breakthrough, such as those by the American Congress on Surveying and Mapping, by Professor Carl Troll and by Professor Vernon Mulchansingh, given in Chapter 1.

PHASE TWO: EXPLORING MAP MESSAGES

A good way to introduce this aspect of your study is by retelling the following story:

> On April 27, 1521, Ferdinand Magellan was killed by Lapu Lapu on the shores of Cebu, Philippines. On the site where Magellan was killed there are now two monuments. One was erected by the Spanish government in 1886 under the reign of Isabella II. The other was put up by the Philippine Historical Society in 1951. The former monument sees history from the point of view of the Spanish Empire, and thus glorifies Magellan as the great leader of the Spanish expansion. The latter, which interprets the same event from the perspective of the Philippine people, commemorates Lapu Lapu as the hero who said no to Western aggression. The color of the Magellan Monument is white and its form is typical of Spanish architectural style. On the Philippine Historical Society monument are painted the faces of Lapu Lapu and his people, as they fought against the invaders. This does not mean necessarily that the Lapu Lapu memorial is artistically superior to the Spanish-style Magellan memorial. But it does point out that the national liberation attained in the postwar period provides the Philipine people with a different perspective from which to see the event.[21]

Turn next to Figure 10. If you are in a classroom setting you may want to show it using an opaque projector; for a small group it may be enough to let all members of the group pass around a copy, either in a second copy of this book or a photocopy prepared in advance. Alternatively, mark up a copy of a Peters wall map using Figure 10 as a model.

Remember the statements made about our traditional maps: that they shortchange the Southern hemisphere in contrast with the industrialized North. This is as true of the maps in our minds as it is of Mercators on our walls and TV screens.

The magnitude of the error may begin to be comprehended as the group ponders the statistics given with the figure.

ROLE PLAYS

Role play will enable your group to enter a quite different dimension of this reality. Against the background of what they have dealt with so far, let them role play one or more of the following situations:

Role Play Number 1 is based on a comment made by a high school student in the Soviet Union when the Donahue Show (NBC-TV) went there in February 1987. You will need two persons: one to be the Soviet student and one to be a person living in the United States or other Western country.

If you prefer, add a third person, to be Phil Donahue, but Donahue should not do more than set the scene and hold the mike for the participants. Let the other two carry the dialogue from there.

The context is a discussion of East-West relationships. Part of that has to do with images: the images we have of one another. The Soviet student, who has studied in America, says: "I think I know why you have such great fear of us. On your maps you always make us look so BIG, so threatening! We're a big country, yes, but not all *that* big!"

Role Play Number 2 is set in India. The number of participants can vary. They are students who have just learned that the map they have used all their lives—and which they always assumed to present the true picture—is a lie. India is not small and inconsequential in relation to the rest of the world, regardless of how it gets depicted. It is, in fact, over three times larger than Scandinavia, for example. Take into account that some of these students still see India depicted in British red; that represents one kind of false information. Now they face a second major question having to do with their self-image. Let them express their reaction. What are some of the questions they would surely ask? How do they feel about all this?

Rules for Role Play are simple but important. The players may be chosen by the leader or by the group or may volunteer. Once chosen, each person must put her/himself as fully as possible into the role. The point is to think, feel, act and talk like that person, not like yourself!

The teacher or leader should "warm up" the players before they appear in front of the group. This may be done where they will not be seen or heard by others in the group. The leader does not tell players how to feel or what to say, but may help them get into their roles by asking leading questions.

Allow a brief time—three or four minutes may be enough—for each role play. Cut before it drags. Allow players to express how they felt about their roles and whether they saw the situation as being realistic.

Next, invite the whole class to comment. Did the situation as presented make you think? Did any ideas come to you along the way? What do you think could be done to work toward a solution? Following an appropriate debriefing or total group discussion it may be valuable to choose other players and redo one or both role plays. Different insights may then emerge.

OPTICAL ILLUSIONS

A different approach altogether to Phase Two picks up from what we all know about optical illusions. An example is shown as Figure 11. You could, in a group, make use of a chalkboard or turnover chart or opaque projector or photocopies to display this or other examples of how the eye—and therefore we—may be deceived.

What we confront in the case of the world maps we carry around inside our heads is very much like that. We look at two equal lines and think that one is bigger. Thus the eye has been fooled, even though—especially for those of us who have

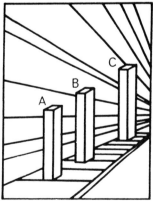

Fig. 11.
Contrary to the illusion created by the background lines, A, B and C are of equal size.

played the game before—we know better. When we see a map the eye is not deceived; it accurately perceives that Scandinavia occupies more map space than India. The eye is not fooled. Worse: The mind is.

So the earth is falsely conceived. We have entered into the process of being deceived. At the far end of that process we may be thoroughly convinced that:

—the North is more important than the South

—we (however the term may be defined) are at the center of things

—we are big; others are small

—nevertheless, the Soviet Union hovers over the world like some terrible giant, massive and threatening

—we are number 1

—we must remain number 1 no matter what

—the rest of the world wants what we have

—we have a mission to the world: to tell them what to believe, how to run their show, how to control their politics, what kind of economic system will be best for them, how much wheat or Pepsi they should buy from us and how many automobiles they should sell to us, etc.

—we are the most advanced country, with the best governmental system, the best ball teams, the best telephones, the best education, the highest literacy rate, the best medical care, the best flag, the best moms, the best apple pie. . . .

You could have a stimulating exchange around just such statements. How true is this? Do group members want to express dissent? What do they find uncomfortable? If we had used the term "brainwashed" instead of "deceived" to label the process described above, would your group's response have been any different? Do participants feel the statements are too strong? Or that they should be even stronger? Do they feel like defending things as they are? If so, why?

PHASE THREE: MAPS THROUGH HISTORY

Space limitations in this handbook have not permitted a treatment of the historical development of the mapmaker's art—now a science—across the centuries. Nevertheless this can be a most revealing and rewarding study. Fortunately, a good treatment is available in *The New Cartography,* especially pages 1–66.

If the wealth of material presented in *The New Cartography*

seems too formidable, we would suggest the following outline:
1. The disc world map of Hecataeus (Hekataios)
2. The rectangular map prepared by Eratosthenes, who calculated the circumference of the earth with an error of less than one percent! Note his introduction of a rectangular grid.
3. The influence of the Romans, who held political and military power.
4. The work of the Greek astronomer Ptolemy, whch continued to be highly influential for centuries.
5. The influence of religious preconceptions: Christian and Musim examples.
6. The dawning of the era of European expansion into the larger world, and how the experiences of that time altered the way people viewed the world and therefore how they did their maps.
7. The contribution of Gerhard Kremer/Mercator.
8. The scientific and historical developments that have rendered the Mercator unacceptable as a general purpose map.

The following questions will help the individual student or the group get into this material:
—What were the forces moving map making forward?
—What forces worked against progress?
—How were the major questions resolved?
—What are the issues presented today as the people of the world seek accuracy and fairness in how their world is depicted?
—What hope do you have, based on the historical record, for the future?
—How can we, as individuals or as a group, contribute to the shaping of that future?

PHASE FOUR: TOWARD A NEW VIEW OF THE WORLD

Open this session by providing each member with a pencil and several sheets of plain paper. Ask them to do pencil drawings of the world as it is perceived or experienced:
—by a baby six months old (hint: the baby knows the limited world of its crib, that there are persons who provide food, who talk to it, etc.)
—the child starting kindergarten
—the 12-year old
—a young person who has access to a car

—an executive of a multinational business company
—a peace and justice activist.

Group members might also be asked to tell how they perceive the world. Do they like to travel? How much opportunity do they have to do so? What impressions do they have of other countries? of other countries' people? How much do they know of the rest of the world? Do they accept the statement often made by overseas visitors to North America that Americans know surprisingly little about the rest of the world? How do they respond to the candid statement made by a Peace Corps Volunteer assigned to Sierra Leone: "I had never heard of Sierra Leone before, and plenty of others at the Peace Corps hadn't either. I mean, most of us had never heard of most of the places where the Peace Corps was going."[22]

Some would claim that it is not only ignorance but arrogance that is our problem. If arrogance exists, in what sense is it fed by seeing/believing the traditional map?

In what sense are people egocentric? In what measure ethnocentric? In what sense are the people of this country caught up in the psychology of "We're number 1!"? Is that a special problem of this country, or is it in some way a human characteristic?

If the baby can enlarge her/his world, pushing the limits beyond the crib bit by bit until he/she becomes, we hope, a fully mature human being with world sympathies and comprehension, can a whole nation grow up? Can a people become more sensitive to the needs, aspirations and cultural integrity of others beyond its borders? Is that part of what we hope for in history? In what sense would such an attitude build the structures of peace, understanding, justice?

Just as participants in your group described the ideal map in Phase One, draw out from them now their picture of an ideal world. It may be useful to consciously coin words to communicate this new vision. For example, if the world we want is not made up of billions of egocentric selves, how would you describe it? Would words like omnicentric or inclusive or fairness-for-all or panecumenical or othercentric or valuecentric or zoecentric (focused on life) begin to convey what you mean?

If the group can come to essential agreement on even the broad shape of such a world, move to the next consideration, which is the question:

What's Holding Us Back?

It would be simplistic to suppose that if we could place a Peters map wherever in the world there is a Mercator-based map, we would have achieved the ideal world. Still, one must ask whether we can ever attain global understanding and justice as long as we cling to invalid understandings, of which the Mercator world is a prime example. Dr. Peters phrased the issue this way:

> The question must now be asked whether the public clings to this false view of the world because they do not realize it is a distortion of the truth or because they wish to delude themselves over Europe's loss of a world-dominant position by retaining the vision of an artificially enlarged Europe manipulated into the center point of the world.[23]

The jury has not yet rendered a verdict with regard to Europe. And while our situation in North America is somewhat different, the central question appears to be very similar indeed; it has to do with our world view, our place in the sun, our centrality on the stage of world history, our pride. On that one also there will be a jury verdict.

Make very specific application of that central question now facing us, if you wish. For example:

• Suppose every classroom in this country, in public, parochial and private schools, put on display a large copy of the Peters world map. What new ways of viewing the world might emerge? Would such a move make it harder or easier for students to learn the facts of geography? Would they be more or less interested in other countries and their peoples? Do you think they might develop a different way of viewing the Third World?

• Suppose TV networks used the Peters map as background for world news broadcasts. Would public interest be generated as a result? What questions would likely emerge? Would the result be positive — immediately? in the long run?

• Suppose the federal government decided to set aside the four-centuries-old map they now use (whether as decoration, whether as a political statement) in favor of the new, equal-area, fair-to-all Peters map. Suppose also that when official statements are made in the field of foreign affairs or treaties are signed or representatives of other governments are welcomed, this map sets the context. What effect might this have? If a map from the days of colonial exploitation sends a message, what message might this new map send?

What Can We Do?

No one can do everything. But everyone can do something. You, whether as an individual or in concert with a group, can have a part in reshaping the world through reshaping people's images of the world. You will know best what you can do and what may prove effective in your situation. The following checklist may be suggestive. Place a check mark beside those ideas that you or your group could consider:

_____ Talk with teachers and curriculum officers of your school system about the maps they use. Be prepared to encourage a change.

_____ Provide copies of this Handbook to decision-makers in education and communications for their comment and response.

_____ Invite group members to display a Peters Projection map prominently in their home for a period of six months. This will serve as a good discussion-starter.

_____ Write or call your local television station to inform them of this new, more accurate world map. Invite their representatives to meet with you to discuss what it could do for their news broadcasts.

_____ Find out who your allies are. For example, churches, peace and justice organizations, schools, librarians, college professors. Talk with them about your interest. Together set attainable objectives.

_____ Ask for a course in your local adult school, community college or similar setting on "Maps and What They Mean" or a related topic of your choosing. This Handbook might serve as the basic course guide, along with other materials on the Resource List at the back of this book.

_____ Develop a story for your local newspaper, school board newsletter or other journal based on this course.

Above all, have the confidence that a new view of the world is not only necessary, it is possible.

The transformation of the world begins in the transforming of our minds. And the renewal of our minds begins with the transforming of the images we entertain: images we hang on our walls and images we carry with us in our heads. Much has been accomplished. Some encouraging things are happening even now. If you are convinced that this new view of the world can contribute positively to the human future on Planet Earth, you have a growing network of support. Now, with your strong help, the future of the world can be just a little brighter!

Notes

1. Peters, Arno, letter to the author.
2. Strictly speaking, the earth is not perfectly spherical, since it bulges somewhat in the middle and is slightly flattened at the poles. The difference is minor, however, so for our present purposes the term sphere is the appropriate one.
3. Peters, Arno, *The New Cartography*. New York: Friendship Press and Klagenfurt, Austria: Universitatsverlag Carinthia, 1983. p. 118.
4. *Ibid.* pp. 69–73.
5. Quoted in *ibid.* p. 58.
6. *Commercial Atlas and Marketing Guide*. Chicago: Rand McNally, 1984. p. F11.
7. A great circle is any line around a sphere that exactly divides the sphere into two halves. The equator is a great circle. All meridians, since they pass through both North and South poles, are also great circles.
8. Most cartographers would claim that an accurate linear scale is not possible overall on any world map. For example: ''. . . it is impossible to maintain a precisely constant linear scale.'' Article, ''Map,'' in *Encyclopaedia Britannica*. Chicago: 1963. Vol. 14, p. 836.
9. A great circle arc is any segment of a great circle. Surprising as it may seem, while a great circle is the *longest* circle that can be drawn around any sphere, a great circle arc is the *shortest* distance between two points on that same great circle. Airlines typically follow great circle arcs.
10. In the popular Friendship Press version one square inch = 158,000 square miles, or 1 cm² = 63,550 km².
11. American Congress on Surveying and Mapping: *Bulletin*. Number 59, Nov. 1977.

It should be noted that the terms ''forms, distances and angles'' in the statement are equivalent to the terms ''shape, linear scale and angle'' used in this handbook. Thus it is clear that the Peters Projection achieves total accuracy of area, fidelity of axis and fidelity of position *as well as* minimal distortion of shape, distance and angle. This combination of qualities on one map had not previously been known.
12. Troll, Carl, *Die Peters-Projektion im Urteil der Fachwelt*. Munchen-Solln: Universum-Verlag, n.d. p. 40.
13. Mulcansingh, Vernon, letter to the author.
14. Gerhard Kremer was a refugee from Flanders during the religious wars of that period.
15. Fuller, Buckminster, *Ideas and Integrities*. New York: Collier Books, 1969. p. 124.
16. Fuller, Buckminster, *Fuller Projective-Transformation* an interpretation of the Dymaxion map.

17. From the interpretation accompanying the map.

18. Trewartha, Glenn T., Robinson, Arthur H. and Hammond, Edwin H., *Physical Elements of Geography*. Fifth Edition. New York: McGraw-Hill, 1967. p. 24. (italics added).

19. *Ibid*.

20. Letter to Friendship Press.

21. Takenaka, Masao, *Christian Art in Asia*. Tokyo: Kyo Bun Kwaan, 1975. p. 18.

22. Coates, Redmon, *Come as You Are: The Peace Corps Story*. New York: Harcourt Brace Jovanovich, 1986. p. 84.

23. Peters, Arno, *the Europe-Centred Character of our Geographical View of the World and Its Correction*. Munchen-Solln: Universum-Verlag, 1979. p.9.

Resources

MAPS

A. The Peters Projection
1 . Popular version, 35″ × 50″. $7.95.
2 . A large wall map suitable for classroom use is planned.
     Usually obtainable through your map store, bookstore, or school supply house. If they do not have Peters Projection maps in stock, write or call:
     Friendship Press Distribution Office
     P.O. Box 37844
     Cincinnati, Ohio 45222-0844.
     Telephone: (513) 948-8733.

B. Other Projections
Other maps mentioned in this handbook may be obtained from their respective publishers: Rand-McNally, Hammond, the National Geographic, U.S. Geological Survey, the Buckminster Fuller Institute.

BOOKS

A. About the Peters Projection
Peters, Arno, *The New Cartography*. New York: Friendship Press; Klagenfurt, Austria: Universitatsverlag Carinthia, 1983. 164 pp., hardcover. In German and English. Commissioned by United Nations University, Tokyo. $20.

Peters, Arno, *Space and Time: Their Equal Representation as an Essential Basis for a Scientific View of the World*. New York: Friendship Press; Klagenfurt, Austria: Universitatsverlag, 1985. Originally presented as a lecture by the author at United Nations University, Cambridge, England. $4.95.

B. General Works
Kidron, Michael and Segal, Ronald, *The New State of the World Atlas*. New York: Simon and Schuster, 1986.

Robinson, Arthur H., *Elements of Cartography*, Second Edition. New York: John Wiley, 1960.

Trewartha, Glenn T., Robinson, Arthur H. and Hammond, Edwin H., *Physical Elements of Geography*, Fifth Edition. New York: Rand McNally, 1967.